Sylvia Böcker

The process of change in the German wind industry

GRIN Publishing

Bibliographic information published by the German National Library:

The German National Library lists this publication in the National Bibliography; detailed bibliographic data are available on the Internet at http://dnb.dnb.de .

Imprint:

Print and binding: Books on Demand GmbH, Norderstedt Germany
ISBN: 978-3-656-31946-7

This book at GRIN:

http://www.grin.com/en/e-book/200258/the-process-of-change-in-the-german-wind-industry

The process of change in the German wind industry

Abstract

The paper identifies the necessity of a deep change process in the German wind industry. The need for change in the wind industry was given by the decision of the German government to finish nuclear energy production in the year 2022. The Fukushima accident lead to this decision and showed dramatically the urgency of finding solutions for a stable and sustainable energy production based on renewable energies. The advantages of using decentralized wind energy which is possible through installation of small wind turbines is so far less acknowledged in politics and economics. The amount of investment and the market share of small wind technology are minimal against their potential and compared to the investments which are put in centralized wind energy production with big 3 bladed turbines. The reduction of the costs for the transmission lines and the electrician transmission losses could be a major economical factor to the point that one could have a parallel grid to fulfil the requirements for power of special regions. To increase the percentage of decentralised produced wind energy it is necessary to identify the key reasons which minimise the success of the innovations in the field of small wind technology. Key factors are f.e. high costs for development and high governmental regulations as well as the lack of suitable technology. Another reason can be seen in the thinking of the core persons and decision makers in this business field as well as in politics. To be able to analyse the complex situation and get an understanding of the interconnection of different levels in this industry the systemic model by David Kantor is used which includes the normally less acknowledged level of mental models in economics.

Key Words

Renewable energy, sustainable energy, Wind industry, Small wind technology, decentralised energy, mental models, systemic model

Introduction

The business field of renewable energy is divided in the wind industry, the solar industry, biomass industry, geothermie and water energy. The German wind industry is one of the major industries in the renewable industries and produces centralised energy, onshore and offshore. It is well known and well accepted worldwide and it is represented by a few big companies. This industry is well described and analysed (Ohlhorst, 2008).

The growing interest in the development of renewable energies in the last 20 years was caused through the discussion of the increasing global warmth and the search for alternatives to fossil energy. Nuclear energy has been so far still an option to ensure stable energy. This changed after the Fukushima accident and the German Government decided to shutdown all nuclear power stations until the year 2022, other countries e.g. Switzerland are following. This changes the requirements for the renewable energy industry in total. The challenge is now to produce cost effective and stable energy for whole nations which is non nuclear and without CO^2. The investments to reach this goal especially for new transmission lines are tremendous and all renewable industries are going to be approved under the new circumstances. Due to these facts it is important to use every option which could widen the chances to reach the given goal and it may include decentralised produced wind energy through small wind technology.

Methods and resources

The aim of this paper is to show the huge possibilities which are given through decentralised small wind technology as so far less acknowledged business and to find out the reasons about the difficulties and the hindering factors in the market. The small wind technology will be seen as one part of the German Wind industry which is one business field in the renewable industry. The change model of Richard Beckhard will be presented as an overview about a change process. The background model for the analysis of the current situation is the systemic model by David Kantor which is known as a team learning model. It will be adjusted to analyse the macro environmental factors in the German wind industry in order to get an insight about the complexity of the whole situation (Kantor, 1994). The market research covered as given in the PESTLE

Analysis economical, environmental, political, social, technical and organisational issues (Hermann, 2002) and was mainly done by a research of literature and in the internet as well as different statistics of the last years. It is divided in the sectors of one the big 3-bladed wind industry and second in the small wind technology sector. First results can be presented and will be discussed and will lead to deeper qualitative research and inquiries which include the mental models of the core persons and decision makers in this business.

Richard Beckhard states that if a process of change is needed the question “why change” has to be answered clearly. Further more you have to get a deep understanding about your current situation and as well as a clear vision or picture of the future situation. Then you can manage the gap between those two situations. To understand the current situation the hard factors, e.g. economical factors as well as the soft factors are to be considered in order to find out which strategies are useful to reach the given goals (Beckhard, 1887)

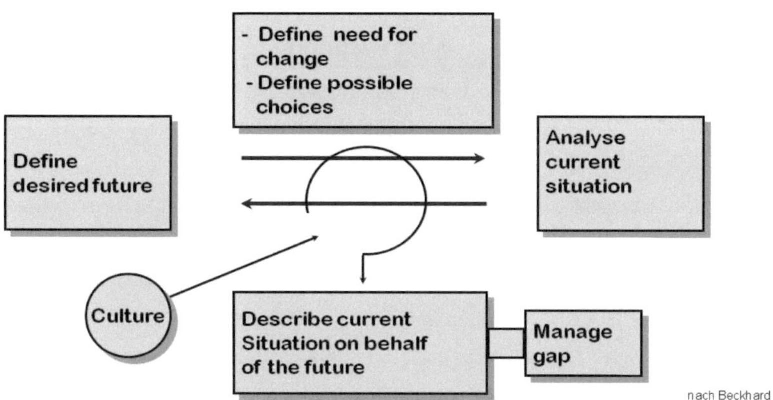

Figure 1 The map of change process by Beckhard 1987

Therein the culture could have a major influence. In this term, culture is not only meant as a national culture but also meant as a professional or an organisational culture which influences the mental models and the deeper beliefs of persons or groups (Schein, 1995).

The systemic model of David Kantor provides an overview over the different structural levels which influence the results in the business and the process and indicates the interconnection between the different structural levels.

Team Learning Model

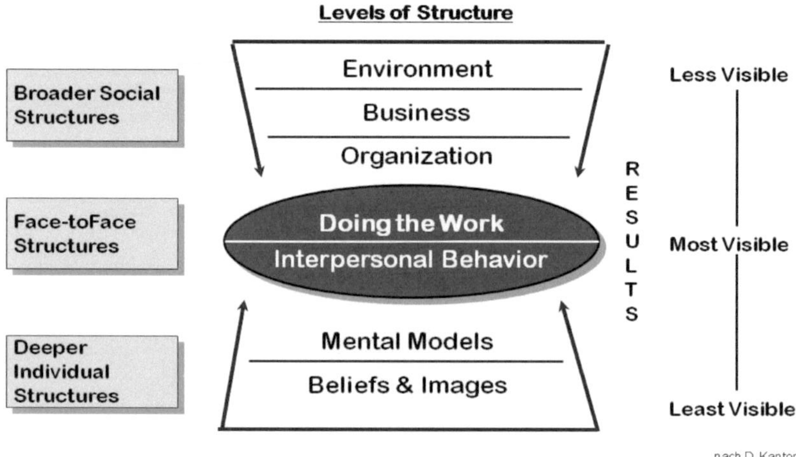

Figure 2 The Team Learning Model by David Kantor,1995

The levels of structure which are well known in the strategic management are the “Broader social structures” which include the environment, the business and the organizational level. Deepika Nath calls these levels “traditional territory” which is, in comparison to the face to face structures less visible (Nath, 2008). The “face to face structures” are the level of obvious interpersonal behaviour, the way people do their job, they way they talk to each other and the way of negotiating and decision making. It is also the level of the visible results, e.g. the governmental decision of the nuclear exit in Germany until the year 2022 is an obvious result of the negotiation process after the

Fukushima accident in Japan. The level of the "deeper individual structures" is the least visible level of the model. It includes the mental models and the beliefs of people as well as the core beliefs, critical images and stories. The level of the deeper individual structures is both personal and cultural (Schein, 1995).

Results

The goal of the change process is defined by the German government as the complete exit of nuclear based energy until the year 2022 including the guarantee of stable and sustainable energy providing which in turn includes a 40% reduction of fossil energy (Bundesregierung.de).

The broader structures of the "German big 3- bladed"wind energy

The main factors in the environmental level are given by the nature of wind. Wind is a natural resource which is hardly to influence. If you produce energy out of the wind you have to consider that you get eight times the energy if you double the wind speed. The wind speed is normally higher offshore and in coastal regions than onshore. Depending on the environment the wind speed increases in the height with a special coefficient which is given by Hellmann. Germany is divided in different areas which are called wind zones. The categories are wind zone 1with less wind, wind zone 2 with more wind and wind zone 3 and 4 with strong wind. 95% of Germany is wind zone 2. Wind zone 3 and 4 are mainly represented offshore and at the coastal lines. Especially in wind zone 2 (inland) it is of importance that the wind speed increases in the height. The power of wind is characterized by the Betz Law which says that the maximum power you can harvest out of the wind is given by 60% of the whole power which is in the wind and touches the area of the turbine. The power coefficient defines the efficiency of the turbines.

The German wind industry is one of the major industries in the field of the renewable industry and is represented by a few companies which produce big 3 – bladed turbines which produce centralised energy. The total turnover of the German industry in the year 2009 covered 6 Billion Euro of the total amount of 33 Billion in the renewable industries. The highest turnover in the renewable industries is reached by the solar industry which is highly funded by the German government and represents a decentralized energy providing because it is mainly installed in the private and commercial sector (BMU, 2010).

In the year 2010 the wind industry provided 6% of the German electricity which is the highest rate of energy providing compared to the other renewable industries (AG Bilanzen,2011).

Percentage of renewable energy in German electricity in the year 2010

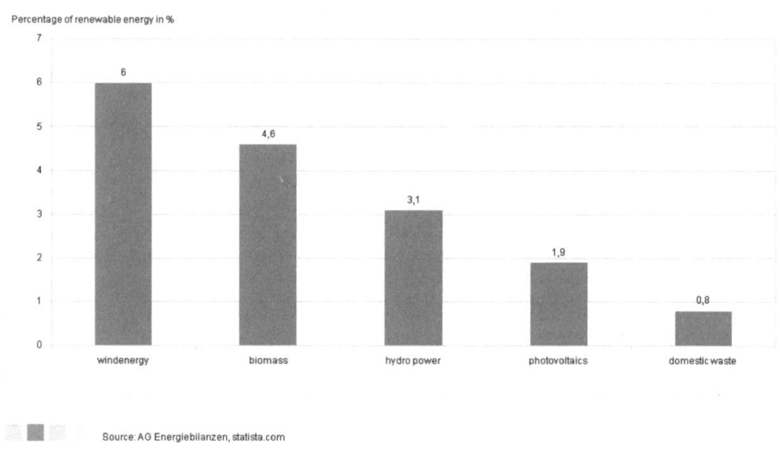

Figure 5 Percentage of renewable energy in Germany in 2010, AG Bilanzen 2011

The German wind industry is well known and well accepted worldwide. The quotation of export in 2009 was about 75% and inland activities about 25% (VDMA, 2010). The Investments in Germany reached an amount of 2,1 Billion Euro in 2010 and worldwide without Germany an amount of 34,1 Billion. The number of installed wind turbines increased from 1652 in the year 1993 up to 21356 in the year 2010 which means an increase of 20 times within 17 years (ISET e.V., 2010). Also the number of employees working in this business field shows the importance of this industry. In the year 2009 the total number of employees was about 140.000, including the employees of the suppliers. The major players are Enercon, Vestas, Nordex, Fuhrländer, Siemens and

Repower systems. Since 1990 a concentration on a few big companies in this sector is observable and since the nineties Enercon and Vestas belong to the biggest players in the market (Ohlhorst, 2008). The increasing influence of the German wind industry was possible through a combination of business and science as well as legislation. Associations and institutes were founded to establish this industry in the society and to expand the knowledge which is also a huge export factor. In the year 1990 the DEWI, the German wind institute, was founded which is one of the most influential and well known institute for measurements, research, analysis and studies as well as prognosis and schooling and consulting worldwide. The ISET "Institute for solar Energy providing technicians" provides scientific measurement- and evaluation programs. Also, the BWE "German Association of wind industry" was founded and as well as the VDMA which is the association for the suppliers (Ohlhorst, 2008). This process went along with several regulations and novels of the EEG (renewable Energy law) which regulates the compensation for electricity fed into the grid for the investors.
The technical development is characterized by an increase of the capacity of the turbines which is mainly reached by higher and bigger turbines. At the same time, this fact is one of the factors for the limits of growth in this industry. Higher turbines need more space and bigger turbines have a high risk to fail, especially in strong wind areas. This was one of the reasons why during the years 2002 – 2008 a phase of consolidation was noticeable. It was remarkable that the space needed for the turbines was limited at the coasts and the trend went onshore which leads to the development of even higher and bigger turbines to reach the profitability. Another reason was that the public began to criticize the big turbines because of their negative influence of health and the basic conflict of interest with nature. The association for environmental protection, on the one hand stated the reduction for the CO^2 emission and, on the other hand, the big turbines were mainly built in areas where the nature is mature and should be protected. Also, a conflict of interest arose with other competing industries like the tourism. These and other reasons influenced the growing interest in offshore wind parks.
Basically the technical design of the horizontal turbines is the 3- bladed style, technical variations are given by the different heights, diameter and the efficiency of the generator and inverter systems.

From the beginning on one of the biggest problems in this business was of how to bring the produced electricity from regions with high wind speeds but limited population to the places where the power is needed. Already in the early years, it was obvious that the grid reached its capacity. A decent storage for wind power which fits the requirements of high efficiency is to be developed but this will take until the year 2030. The association of the German banks states in their position paper of October 2011 that the energy transmission will cost in total 200 Billion Euros being able to ensure a stable energy production with a reduction of 40% CO^2 emissions (Bankenverband, 2011). The costs for the necessary expansion of the grid of 3600 KM will be 10-50 Billion depending on the chosen material. This calculation does not include the costs for the losses of maximal 153000 Euro per kilometer annually and the costs for the maintenance of 3000 Euro per kilometer annually. The consequences for the environment are not respected in this calculation (DENA, 2011).

The broader structures of small wind technology

The small wind technology is an approach which focuses on decentralization of energy production. Through a long period of time, the small wind technology was of less interest but in the last years the interest of the public in decentralized produced energy through wind starts to be increasing. Although it is also an industry in the wind sector, the characteristics are different in comparison to the wind industry with the big turbines. The small wind technology gives the possibility to harvest the energy out of the wind where it is needed and can be built in the current infrastructure. It can be used as street lighting, it can be put on roofs and under bridges if you have the right technical design.
The small wind technology is mainly represented by small- and middle sized companies which joined together and, in the year 2008, founded the "German association for small wind turbines BVKW" to grasp ground and to get more political influence in order to establish better conditions for this type of wind technology (Otter, 2009). This association was also founded because in the "German wind association" there was no interest to include small wind turbines (Kroeger, 2010). The BVKW differentiates between micro turbines, turbines for houses, turbines for an autarkic energy system and turbines up to a rotor area of 200 m^2.

The trend worldwide shows an increase of installed capacity in the small wind sector from 25 Megawatt in the year 2007 up to 43 Megawatt in the year 2009 as shown in figure 4.

Capacity of Small Wind Turbines since 2007

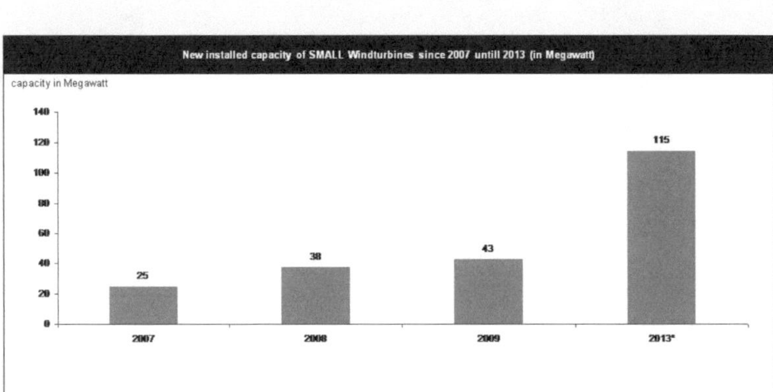

Source: Handelsblatt, Handelsblatt, No.: 182, 21. September 2010, page 50

Figure 4 Trend and forecast small wind turbines worldwide 2007-2013, Handelsblatt 2010

The forecast was given until the year 2013 with 115 Megawatt. The forecast was calculated before Fukushima.

Germany, which is one of the market leaders in big wind technology worldwide, has a minimal percentage of market shares, only 3% in the small wind sector, in comparison to the USA which holds an amount of 40%, Great Britain which holds an amount of 25% and the rest of the world which holds an amount of 32% (Handelsblatt, 2010) which is described in figure 5.

Market shares in percentage of SMALL Wind Turbines worldwide

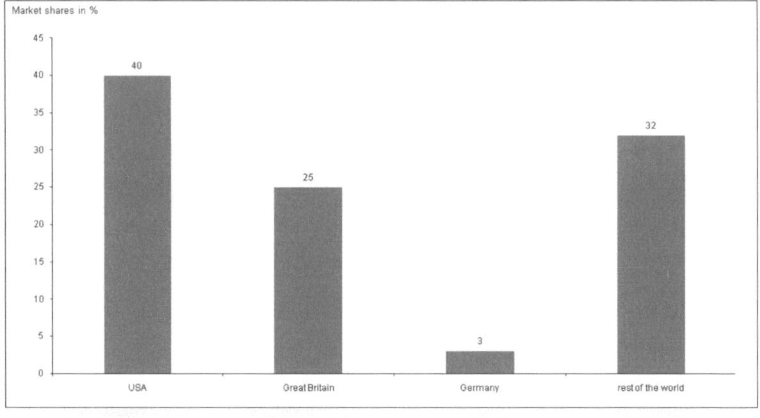

Source: Handelsblatt, Handelsblatt, No.: 182, 21. September 2010, page 50

Figure 5 Market shares of Small wind turbines 2010, Handelsblatt 2010

One reason could be that f. e. Great Britain changed its regulation and provided external conditions for the support for installation of decentralized wind energy. The fed in tariff was increased, similar to the German solar industry.

The regulations to get the permission for the installation of a small wind turbine are similar to the given regulations of big wind turbines although they are used to have a height of 20-30 m up to 50 m. The process is cost intensive and long lasting. A general regulation in Germany is not to be defined in the near future, the regional differences is the current state (Otter, 2009).

The German BVKW claims a unified regulation and, furthermore a minimization of the requirements to open the market more effectively for small wind turbines, also the fed in regulations should be adjusted.

In contradiction to the big turbine versions the small wind technology provides numerous variations of technical designs. Most of them are similar to the horizontal 3-bladed versions, but also 2-bladed turbines are available. Other versions are the Darrieus or the multiblade American turbine type.

A big difference in the technical design is provided by the vertical turbines which are characterized as the Savonius type. The Savonius type is the variation which was first patented in 1927 and makes a big difference to all other turbines. It is the only one which works without noise and has less vibration and can be installed in the infrastructure. In the last time, the interest for vertical turbines is increasing because of its possible usage in the infrastructure. Vertical turbines can be installed where other turbines are forbidden because of their noise and their frequencies and the produced shadows. The potential of the Savonius style is well known in the scientific level and a lot of investigations had taken place (Komatinovic, 2006).

To be able to harvest the most power out of a vertical turbine a lot of research has to be made because of the lack of suitable supporting systems like generators and inverters. It is of crucial importance for the wind system that it can work in its optimal "Tip Speed Ratio" which is only assured through a generator and inverter system that suits to the individual characteristics of the turbine, otherwise its efficiency is minimized. To find out the optimal "Tip Speed Ratio" and the individual power curve, a lot of measurements have to be done which is highly cost intensive. Another problem is the fact that once the optimal tip speed ratio is known, the suiting technology has to be developed in itself, especially for the vertical versions because the most generators which are available in the market suit the characteristics of the horizontal turbines.

Due to the fact that the development process is long lasting and cost intensive, the products on the market are often still in the innovation process and do not reach their marketability or the determination of the given products is not fulfilled and a lot of promises are not reliable. This minimizes the trust of the customer in small wind technology and this is the reason why the BVKW claims for unified definitions of the turbines through certification. The certification process itself is also long lasting and cost intensive and adopted by the 3- bladed technology.

To show the potential of the Savonius small wind technology you have to think in big numbers, e.g. one vertical turbine in wind zone 2 could produce 15.000 KWh and 100 of them 1.500 000 KWh which 1.500 MWh. This means that it is also possible to ensure an energy solution for special regions which is independent from the grid.

Discussion

The results of the analysis of the broader structures show the difficulties the small wind technology is facing to enter the market. The innovation processes are cost intensive and long lasting. The German wind industry is divided in two sectors which have a complete different approach. The results in the given negotiation process of finding solutions for the energy transition demonstrate that the decentralized approach which is represented by small wind technology is less acknowledged because the solutions focus on centralized energy. The increase of offshore wind parks and the building of huge transmission lines are the most important solutions at this moment of time to reach the goal for the nuclear exit. One important reason is that the small wind technology has less influence due to a lack of reliable products which is given through the difficulties of the market entry and expansion. This shows a vicious circle. This dynamic is increasing because the knowledge about the characteristics of wind which is the basis to find technical solutions is covered by the 3 – bladed industry and the associations and institutes who are working in this sector. The small- and middle sized companies are often not able to pay the intensive measurements and innovations for the necessary additional equipment. To get public founding for projects like this the people in charge have to be convinced which is also difficult because of the lack of well working products in small wind technology and the lack of alternative vertical turbines. It seems that the fact that Germany is one of the market leaders in wind industry worldwide is a hindering factor for the innovation processes in this field. Other countries, e.g. Great Britain, do not have such a big industry with the big 3- bladed turbines and opened up for new strategies. The worldwide trend shows that a combination between decentralized and centralized produced energy is needed to fulfill the requirements of the future. One consequence in a long term range could be that the German wind industry runs the risk to minimize their worldwide influence in this business field. The minimal percentage of market share in small wind technology could be interpreted as one of the first symptoms.

It is also obvious in the given data that the decentralization of wind energy could reach a high amount of turnover if the regulations could be changed as in the solar industry. The amount of produced energy could unburden the strained grids and strengthen the innovation processes.

Conclusion

Further research is needed to understand the thinking behind the decisions and to help to reframe the mental models.

Summary

The huge challenge of a transition process which includes the whole energy sector in Germany given by the decision of the German government involves a use of all available resources. No technical options and natural potentials must be left out. The survey shows the importance of a deep analysis of all involved industries including all technical possible innovations. A limitation through one way thinking processes of the core persons could minimize the quality of the results and will cost wasted time and money and runs the risk of not achieving the goal of the nuclear exit. The objective was to visualize the potential of small wind technology and to find out hindering factors for the expansion into the market. This was given by an analysis of the broader social structures. Key factors are the lack of money for decent research and the technical finalization to reach the market entry. The results in the energy transition process show the interconnection between the different structural levels given in the Kantor analysis model. The current dominant mental model of people involved in the decision making process is illustrated by taken decisions. The current regulations are ineffective and do not suit the small wind sector. Under these conditions the high potential of decentralized wind energy cannot be used to minimize the cost for transmission lines. A further research is needed to dig into the mental models and to find out strategies to widen the thinking processes.

References

monographs

BECKHARD DICK, REUBEB T.HARRIS, 1987: Organizational Transitions: Managing Complex change, 10f., Addison-Wesley Pub.Co, University of Michigan, ISBN 0201108879

BRUNS ELKE, KÖPPEL JOHANN,OHLHORST DÖRTE, SCHÖN SUSANNE, 2008: Die Innovationsbiographie der Windenergie, Absichten und Wirkungen von Steuerungsimpulsen, 28-62, LIT Verlag Dr. W. Hopf, Berlin, ISBN 978-3-8258-1625-4

HERMANN SIMON,VON DER GATHEN ANREAS, 2002, Das grosse Handbuch der Strategieinstrumente, 199-265, Campus Verlag, Frankfurt am Main, ISBN 978-3-593-39335-3

KANTOR DAVID, 1994: Reframing Team Relationships. In SENGE Peter M.(ed) The fifth discipline Fieldbook. New York: Doubleday, 407 f., ISBN 0-385-47256-0

KOMATINOVIC NEMANJA, 2006, Investigation of the Savonius-type Magnus Wind turbine, Master Thesis, 14-15, Technical University of Denmark,

OHLHORST DÖRTE, 2009, Windenergie in Deutschland, Konstellationen, Dynamiken und Regulierungspotenziale im Innovationsprozess, Dissertation Freie Universität Berlin, Verlag für Sozialwissenschaften, Wiesbaden, 184-202, ISBN 978-3-531-16841-8

OTTER PHILIPP, PEHNT MARTIN, 2009, Pilotstudie zur Akzeptanz vertikaler Windenergieanlagen, 4f., ifeu- Institut für Energie- und Umweltforschung Heidelberg GmbH

SCHEIN EDGAR H., 1995, Unternehmenskultur, Ein Handbuch für Führungskräfte, 20-30, Campus Verlag ISBN 3-593-35268-0

NATH DEEPIA, Building trust and cohesiveness in leadership Team. IN: Reflection Volume 9, Number 1, 24-35, The SOL Journal, ISBN 1524-1734

Studies/Internet/Others

Der Weg zur Energie in der Zukunft- sicher bezahlbar und umweltfreundlich, www. bundesregierung.de, Energieeckpunkte of June 6th 2011

BUNDESVERBAND DEUTSCHER BANKEN, Positionspapier des Bankenverbandes zur Finanzierung der Energiewende, Oktober 2011

DENA, dena Netzstudie – Integration erneuerbarer Energien in die deutsche Stromversorgung im Zeitraum 2015-2020 mit Ausblick 2025, Zusammenfassung der wesentlichen Ergebnisse der Projektsteuerungsgruppe, 2011

KROEGER M, CEO, BVKW Bundesverband für Kleinwindanlagen, telephone conference, July 2010

Statistics

Anteile der Hersteller an neu installierter Leitung durch Windenergie, Deutschland, Deutsches Windenergie Institut DEWI 2010, published DEWI 01. February 2011, statista.com

Beschäftigte in der deutschen Windindustrie seit 2004, Bundesverband der Windenergie e.V., DEWI 2004-2009, FAZ Nr.192192/2010 P.17 Frankfurter Allgemeine Zeitung, statista.com

Anzahl der Windkraftanlagen in Deutschland seit 1993, Institut für Solare Energieversorgungstechnik ISET, 1993-2010, ISET e.V. 12/2010, statista.com

Struktur der Bruttostromerzeugung durch erneuerbare Energien in Deutschland, AG Bilanzen, 2010, AG Energiebilanzen 20.01.2011, statista.com

Neu installierte Kleinwindanlagen weltweit von 2007 bis 2013(Megawatt), Handelsblatt 2010, statista.com

Umsatz mit erneuerbaren Energien in Deutschland 2009, BMU-KI III 1 Zentrum für Sonnenenergie und Wasserstoff Forschung Baden Württemberg (ZSW) 2009, BMU 18.03.2010, statista.com

Marktanteile bei Kleinwindanlagen weltweit 2010 (in Prozent), Handelsblatt, 2010, statista.com